What Does It

Our eyes help us find out t
shape, size, and position of something.

. Is the truck gray or white?

. Which toy looks the biggest?

. Which toy looks farther away than the others?

Draw the shape of the ball.

What Does It Sound Like?

Our ears help us find out what something sounds like and where it is. Sounds can be loud or soft. They can be high or low.

A shout is loud. A whisper is soft. Color the thing that makes the louder sound.

A bird's sound is high. A lion's sound is low. Color the thing that makes the lower sound.

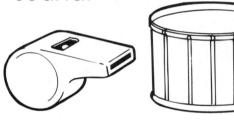

Anna hears a high sound on her right. Color the thing that is making the sound.

What Does It Smell Like?

Our nose helps us find out what something smells like. Smells travel through the air.

Some things smell good. Some things smell bad. Color the things that smell good to you.

Draw something else that smells good to you.

Draw something else that smells bad to you.

What Does It Feel Like?

Our skin helps us find out what something feels like. We usually touch things with our hands.

Choose and write the word that says how each thing might feel.

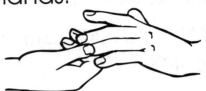

hot
cold

_____

rough
smooth

_____

soft
hard

_____

hot
cold

_____

soft
hard

_____

rough
smooth

_____

What Does It Taste Like?

Your tongue helps us find out what something tastes like. Foods have different tastes.

Write **sweet**, **sour**, or **salty** to tell how each food tastes.

_____   _____   _____

_____   _____   _____

Draw two more foods and write how they taste.

_____   _____

Using Our Senses

We use our senses to observe the world around us.

Look at your pencil. Draw and color a picture of it.

Smell your pencil. Does it smell like wood, dirt, or something else?

What sound does your pencil make when you write with it?

Touch your pencil. Write how it feels.

Observing With Our Senses

Get a food item to observe (apple, sandwich, cracker, candy bar, or other food). Draw it in the lunchbox. Write what you observe.

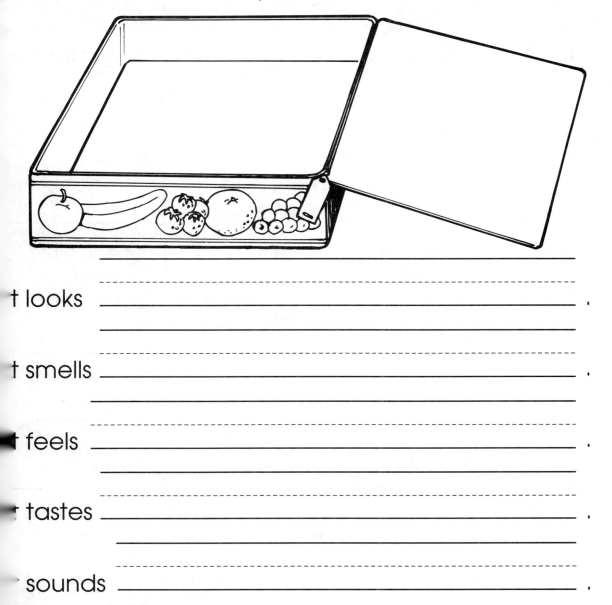

It looks _____.

It smells _____.

It feels _____.

It tastes _____.

It sounds _____.

We Go Together!

Write another word that goes with each group. Use the words in the box to help you.

1. **Weather**
   windy
   cloudy
   sunny

   _____

   vine
   star
   smelling
   whale
   rainy

2. **Animals**
   duck
   wolf
   ant

   _____

3. **Senses**
   seeing
   hearing
   tasting

   _____

4. **Plants**
   tree
   bush
   grass

   _____

5. **Space**
   moon
   sun
   planet

   _____

Describe It!

Write two words from the box to describe each picture. You may use a word more than once.

big	small	soft	bumpy
sour	rough	sweet	round

lemon

kitten

rhinoceros

pineapple

Living and Nonliving Things

Living things grow. Plants, animals, and people are living things. Color the living things.

Living Things Have Needs

Living things need food, water, air, and a place to live. Read each clue. Draw a line to the matching living thing.

1. My home is the desert, where the air is clean. I get food and water from desert plants.

aspen tree

2. I live in the mountains. Rain waters my roots. Sunlight helps me make my own food.

cocker spaniel

3. I make my home with a human family. They feed me two times a day. They always leave me a bowl of water. I sleep on a soft rug in the kitchen.

desert tortoise

Plant or Animal?

Plants and animals are living things. Most plants make their own food. Animals get food from their surroundings. Most plants stay in one place. Most animals move on their own.

Pretend that the plant and the animal below are telling each other about themselves. Write what they are saying. Use the facts above.

Find the Plants

Color the plants in the picture.

Plants' Needs

Plants need water and light to grow.

This plant hasn't had water or light. Color the drooping plant green with some yellow leaves. 	This plant has had light but no water. Color the drooping plant green with some brown leaves. Draw the sun shining. 
This plant has had water but no light. Color the plant green with some yellow leaves. Draw a watering can. 	This plant has had water and light. Color the healthy plant green. Draw the sun and a watering can. 

Parts of a Flowering Plant

Color the flowering plant. Label the plant parts. Use the words in the box.

bean plant

flower	leaf	seed
fruit	root	stem

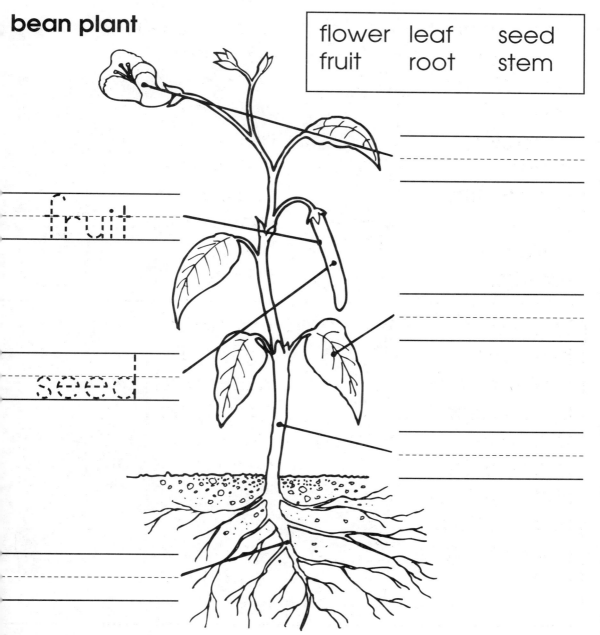

fruit

seed

Roots, Stems, and Leaves

Write **roots**, **stem**, or **leaves** in each blank. Draw a line to the matching part of the plant.

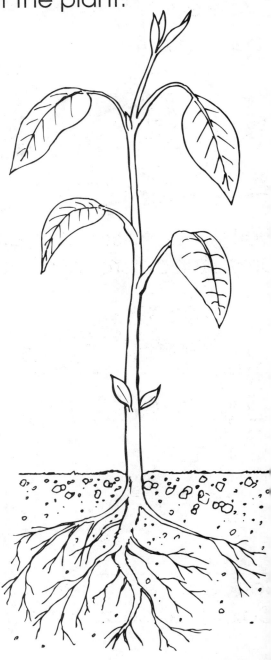

1. The _____ of a plant make food for the plant. They need light, water, and air to do their work.

2. The _____ of a plant holds the plant up. It also carries water, minerals, and food to other parts of the plant.

3. The _____ of a plant take in water and minerals. They are usually underground.

Flowers, Fruits, and Seeds

Flowers make seeds.
Circle each flower on the page.

Fruits or pods sometimes cover the seeds.
Draw a box around each fruit or pod on the page.

Seeds grow into new plants.
Color the seeds in the fruits that you can see.

Eating Plant Parts

We eat plant parts every day. Circle each plant part that we eat. Then write the name of the part or parts.

| flower | leaves | seeds |
| fruit | root | stem |

tomatoes
_____ _____

lettuce
_____

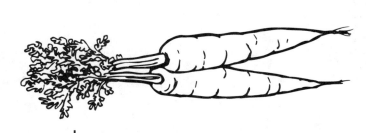

carrot
_____

broccoli
_____

Using Plants

Many things we use are made from plants. Draw a line from each picture to the plant it came from.

oak tree

aloe vera plant

maple tree

cotton plant

What Is an Animal?

Draw a line from each picture to the matching animal fact.

Fact 1: All animals need food, water, air, and a place to live.

Fact 2: Most animals can move on their own.

Fact 3: Most animals have babies that will look like they do.

Draw your favorite animal. Write its name.

© Frank Schaffer Publications, Inc.

Group the Animals

Put the animals in two groups by size.

Large Animals	Small Animals
_____	_____
_____	_____
_____	_____
_____	_____
_____	_____

In what other way could you group the animals?
_____

Home Sweet Home

Animals have many different kinds of homes.

1. The hermit crab looks for a used shell that is just the right size. Circle the best shell for this hermit crab.

2. The weaverbird weaves grass around itself to make a ball-shaped nest. Color the weaverbird's nest light brown.

3. Some rabbits dig tunnels that have rooms. Sometimes many rabbits live together. A mother rabbit digs a different tunnel for her babies. Draw some baby rabbits in their home.

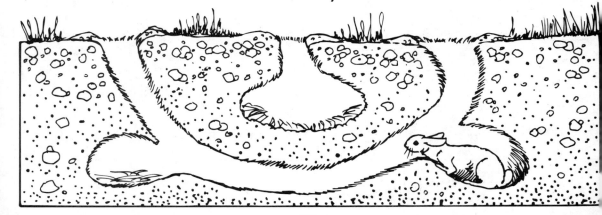

Let's Eat!

Animals eat plants, other animals, or both.
Circle in green the animals that eat plants.
Circle in red the animals that eat other animals.

snake

caterpillar

lion

cow

chameleon

giraffe

On the Move

Animals move in different ways. Write the word from the box that tells how each animal moves. One word can be used twice.

crawls
flies
hops
runs
swims

frog
_____

cheetah
_____

snail
_____

seal
_____

inchworm
_____

hawk
_____

Animals and Their Babies

Pull out pages 25–28. Cut out the two-sided cards. Play **Name That Baby** and **Animal Shuffle**.

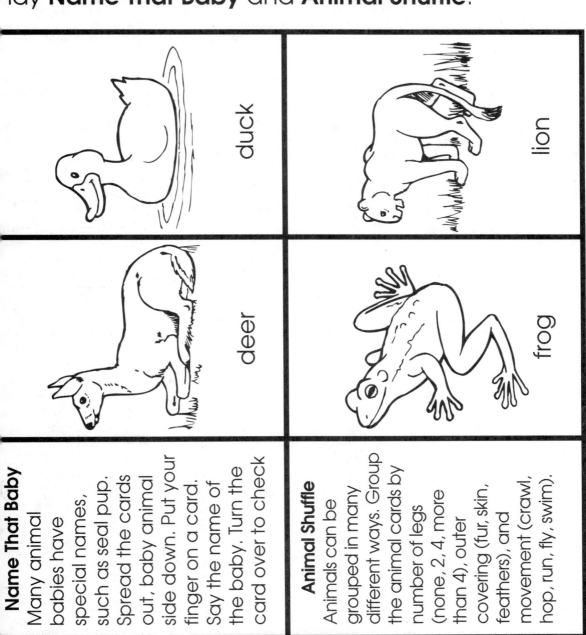

Name That Baby
Many animal babies have special names, such as seal pup. Spread the cards out, baby animal side down. Put your finger on a card. Say the name of the baby. Turn the card over to check.

Animal Shuffle
Animals can be grouped in many different ways. Group the animal cards by number of legs (none, 2, 4, more than 4), outer covering (fur, skin, feathers), and movement (crawl, hop, run, fly, swim).

Animals and Their Babies

Pull out pages 25–28. Cut out the two-sided cards. Play **Name That Baby** and **Animal Shuffle**.

lion cub

duckling

tadpole

fawn

You can use the adult or the baby side of the cards. Then think of your own ways to group them. Show someone one of the groupings you thought of. Let him or her guess how you grouped the animals.

your answer. If you are right, keep the card. If you are wrong, turn the card back over and wait two turns to pick it again. See how many cards you can collect.

Pull-Out Answers

Page 1
1. gray
2. the robot
3. the teddy bear

The child should draw a circle.

Page 2
top left box—lion
top right box—drum
bottom box—bird

Page 3
Answers will vary.

Page 4
hot   smooth
soft   cold
hard   rough

Page 5
salty sour   sweet
sweet salty sweet

bottom box—Answers will vary. The child should draw two more foods and write how they taste.

Page 6
Answers will vary.

Page 7
Answers will vary.

Page 8
1. rainy   2. whale
3. smelling 4. vine
5. star

Page 9
Answers may vary. Possible answers: lemon—sour, small, round; kitten—small, soft, sweet; rhinoceros—big, rough, bumpy; pineapple—rough, sweet, bumpy

Page 10
These should be colored: tree, flower, corn plant, lizard, girl, woman, starfish, sheep, butterfly.

Page 11
1. desert tortoise
2. aspen tree
3. cocker spaniel

Page 12
Answers will vary.

Page 14
top left box—Plant should be colored green with some yellow leaves.
top right box—Plant should be colored green with some brown leaves, sun drawn.
bottom left box—Plant should be colored green with some yellow leaves. A watering can should be drawn.
bottom right box—Plant should be colored green. A sun and a watering can should be drawn.

Page 15

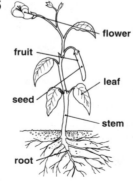

Page 16
1. leaves
2. stem
3. roots

Page 17
Four flowers should be circled. A box should be drawn around the pea pod, orange, peach, peanut, and watermelon. The seeds in the orange, peanut, watermelon, and peach should be colored.

Page 18
tomato—fruit, seeds
lettuce—leaves
carrot—root
broccoli—flower, stem

Page 19

Page 20
roadrunner—Fact 2
beaver—Fact 1
koalas—Fact 3

Page 21
large animals—elephant, whale, alligator; small animals—robin, butterfly, bat; other ways of grouping—Answers will vary. Suggested answers: number of legs, flying/non-flying, mammals/non-mammals, land/water/air animals, fur/feathers/scales, meat-eaters/plant-eaters

Pull-Out Answers

Page 22
1. The third shell should be circled.
2. The first nest should be colored light brown.
3. Baby rabbits should be drawn in the middle tunnel.

Page 23
circled in green—caterpillar, cow, giraffe; circled in red—lion, snake, chameleon

Page 24
frog—hops; cheetah—runs; snail—crawls; seal—swims; inchworm—crawls; hawk—flies

Pages 25–28
The child can pull out pages 25–28, cut out the cards, and play the two games.

Page 29
Answers will vary.

Page 30
Answers will vary. Suggested answers:
Trees and giraffes are tall; they grow; they need air and water.
Trees don't move; giraffes move. Trees are green and brown; giraffes are brown. Trees make their own food; giraffes eat leaves.
Giraffes and people both move; both have to find food; we both grow; we both need air and water.
Giraffes live outside; people live inside. Giraffes have more fur than people do. Giraffes are much taller than people.

Page 31
1, 3, 4, 2

Page 33
Answers will vary.

Page 34
ice, water, steam

Page 35
1. no
2. yes
3. move the paper
4. Answers will vary.
5. air

Page 36
Answers will vary somewhat, but suggestions follow:

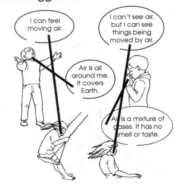

Page 37
windy   snowy
cloudy sunny
rainy   sunny

Page 38
The child should color the children in boxes 1, 3, 4, and 6.

Page 39
Answers will vary. The child should observe the weather for five days and fill in the chart. Then the child can tally the different kinds of weather.

Page 43
1. warm
2. yes
3. T-shirt
4. Accept any answer between 40°F and 60°F.

Page 44
Answers will vary.

Page 45
1. B   2. E   3. C
4. F   5. A   6. D

Page 46
Layers from outermost to innermost: crust, mantle, outer core, inner core

Page 47
Answers will vary.

Page 48
These should be crossed out:
(lever) stairs,
(pulley) wagon,
(wheel and axle) knife.

Page 49
These should be crossed out:
(inclined plane) wrench,
(wedge) crane,
(screw) tweezers.

Page 50
Answers will vary. The child should color the foods he or she likes.
1. Answers will vary. Accept any two fruits.
2. Bread, cereal, rice, and pasta group

Homework Helper Record

Color a raindrop for each page you complete.

Animals and Their Babies

Pull out pages 25–28. Cut out the two-sided cards.
Play **Name That Baby** and **Animal Shuffle**.

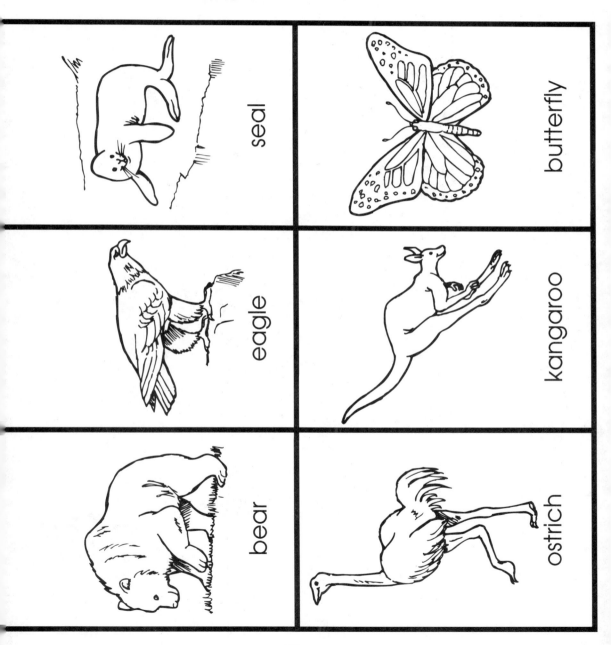

Animals and Their Babies

Pull out pages 25–28. Cut out the two-sided cards. Play **Name That Baby** and **Animal Shuffle**.

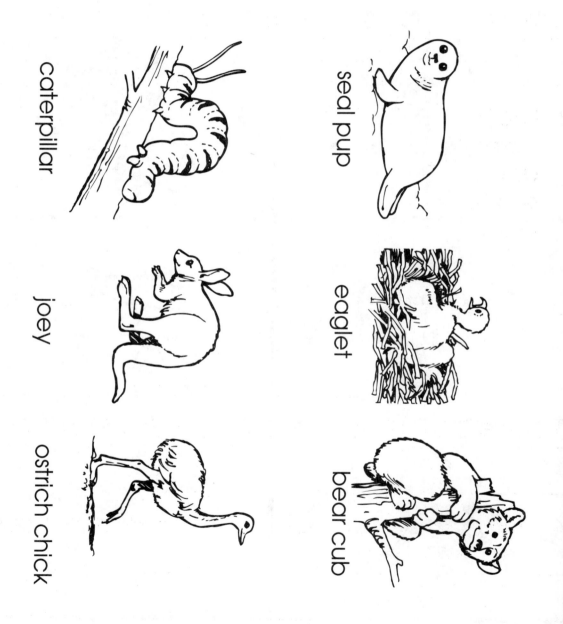

caterpillar

seal pup

joey

eaglet

ostrich chick

bear cub

© Frank Schaffer Publications, Inc.

FS111159 Science

Observing an Animal

Find an animal to observe. It can be a pet, an insect, a wild bird, or any other animal.

1. Write the name of the animal.

_____

_____

2. Look carefully at the animal. Draw it.

3. Write three words that tell about the animal.

_____

4. How does the animal move? (walks, runs, hops, flies, swims, crawls)

_____

Like a Giraffe?

List ways a tree is like a giraffe.

List ways a tree is different from a giraffe.

List ways you are like a giraffe.

List ways you are different from a giraffe.

What Is Water Like?

Write the number of the picture next to the water fact it shows.

____ Water looks clear.

____ Water feels wet.

____ Water can be hot or cold.

____ Water takes the shape of its container.

Water, Please

Plants, animals, and people need water to live.

Draw rain to water the tree and grass.
Draw a pond for the raccoon to drink water from.
Draw glasses of water for the children to drink.

Using Water

We use water to drink and in many other ways.
Color the water. Circle how you used water today.

for fun

for cooking

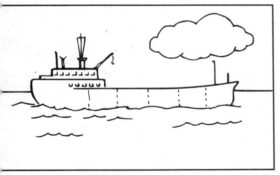
to move things by boat

to make electricity

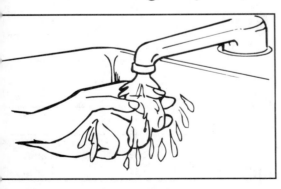
for washing

for watering plants

Ice, Water, and Steam

Water is the only thing on Earth that is found naturally in three forms.

Water that you pour into a cup is a **liquid**. Draw a cup of water.	Water in your freezer turns into ice—a **solid**. Draw an ice cube.
Water that is boiled on your stove turns into steam—a **gas**. Draw steam coming from the teapot. 	Write what water is called as a solid, a liquid, and a gas. solid _____ liquid _____ gas _____

Air Mysteries

Answer the questions to learn the mysteries of air.

1. Air is all around you, but do you see it?
   _____
   _____

2. Fold a sheet of paper to make a fan. Fan yourself. Did you feel the moving air?
   _____
   _____

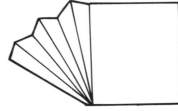

3. Fan another sheet of paper. What did you see the moving air do?
   _____
   _____

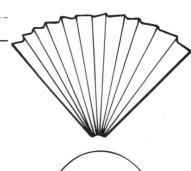

4. Air takes up space, but can you see it taking up space?
   _____
   _____

5. Look at the balloon. What is taking up space inside it?
   _____
   _____

© Frank Schaffer Publications, Inc.   FS111159 Science

What Is Air Like?

Make the speech bubble point to the person who said each air fact.

Weather Words

Weather is what the air is like outside. Write the word from the box that tells about the weather in each picture. Use one word two times.

| sunny | cloudy | windy | rainy | snowy |

Weather Wear

Who is dressed for the weather? Color each child whose clothing is right for the weather.

Weather Log

Keep track of the weather for five days. Draw one or more of these symbols for each day.

sunny   cloudy   windy   rainy   snowy

Month					
Day	Monday	Tuesday	Wednesday	Thursday	Friday
Weather					

How many days were . . . .

sunny? _____   cloudy? _____   windy? _____   rainy? _____   snowy? _____

Kinds of Clouds

There are three main kinds of clouds.

Cumulus clouds are white, fluffy, and start low to the ground. They bring fair weather.

Cirrus clouds are white, wispy, and high. They bring fair weather.

Stratus clouds are gray, layered, and very low. They may bring rain or snow.

Draw and color a scene with cumulus, cirrus, or stratus clouds in the sky.

A Windsock

Make a windsock to measure the wind. You will need paper, scissors, string, tape, and a pencil.

1. Lay a sheet of paper horizontally.
2. Cut slits along the bottom.
3. Roll the paper into a tube. Tape it.
4. Poke three holes along the edge that does not have slits.
5. Thread three pieces of string through the holes and tie.
6. Tie the loose strings together.

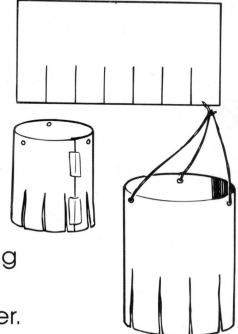

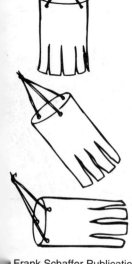

Now go outside. Hold the strings of the windsock out in front of you. Watch what it does.

If it hangs down, it is a calm day.

If it flutters, it is a breezy day.

If it blows sideways, it is a windy day.

The Water Cycle

Read and trace.

1. The sun heats Earth's waters. Heat changes water to water vapor, a gas. Water vapor rises. This is called _____ evaporation.

2. As water vapor cools, it forms clouds. This is called _____ condensation.

3. Clouds become heavy with water droplets. Water falls to the ground as rain. This is called _____ precipitation.

Hot or Cold?

The air can be hot, warm, cool, or cold. We measure air temperature in degrees with a thermometer.

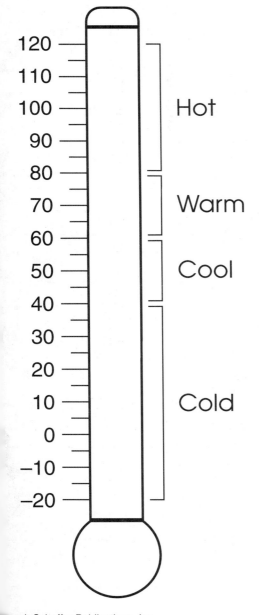

1. Amy said it is 70°F. Is it warm or cold?
   _____
   _____

2. Shane heard it will be 30°F tonight. Should he wear a jacket?
   _____
   _____

3. Should Josie wear a T-shirt or a sweater when it is 90°F?
   _____
   _____

4. Tommy said it was cool yesterday. What might the temperature have been?
   _____
   _____

Weather Thoughts

Draw a weather symbol in each box. Finish each sentence.

sunny

cloudy

windy

rainy

snowy

1. What I like about  weather is
   _____
   _____

   What I don't like about it is
   _____
   _____

2. When it is 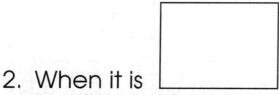, I like to
   _____
   _____

   but I don't like to
   _____
   _____

All About Soil

Match the pictures to the sentences to learn about soil.

1. The land is made up of rocks and soil.

2. Soil is broken down rock mixed with plant and animal matter.

3. Earthworms dig tunnels that let air and water in soil.

4. Plants need good soil to grow.

5. Farmers need good soil for growing food.

6. Water causes soil to settle into layers.

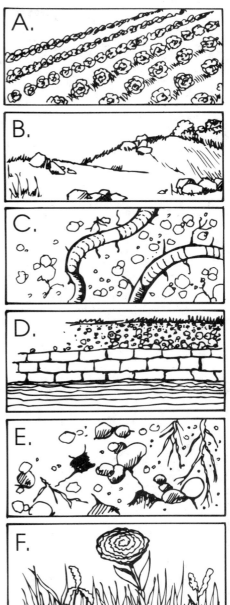

Rocky Earth

Did you know that Earth has layers? They are all made of rock. Read each description. Label the Earth's layers using the bold words. Color the layers different colors.

The **inner core** is solid rock even though it is very, very hot. The weight of the rocks on top of it makes it solid.

Under the mantle is the very hot **outer core**. The rock there is liquid.

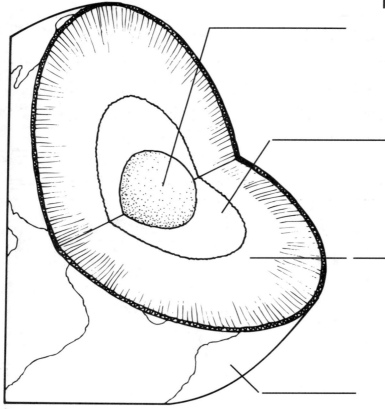

The **mantle** is under the crust. Some of its rock is part liquid and part solid.

We live on Earth's top layer—the **crust**. It is solid rock.

Observe a Rock

1. Find a rock to observe. You will need a penny, too.

2. Where did you find your rock?
   _____
   _____

3. Look closely. What colors do you see?
   _____
   _____

4. What shape is the rock? Is it smooth or rough?
   _____
   _____

5. Scratch it with your fingernail and then with a penny to see how hard it is. What happened?
   _____
   _____

6. Trace the rock here.

Frank Schaffer Publications, Inc.   FS111159 Science

Simple Machines

Read about each simple machine. Then cross out the picture that is not an example of the machine

A **lever** is a bar for raising something heavy. A wheelbarrow and a seesaw are levers.

A **pulley** is a grooved wheel that holds a rope. It helps move things. A tow truck and a flagpole have pulleys.

A **wheel and axle** makes it easier to push or pull something. A car and a wheelchair have wheels and axles.

More Simple Machines

Read about each simple machine. Then cross out the picture that is not an example of the machine.

An **inclined plane** is a slanting surface connecting two levels. It helps move things up and down. A slide and a ramp are inclined planes.

A **wedge** is a tool with a slanting side that ends in a sharp edge. A knife and an ax are wedges.

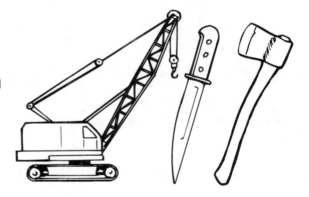

A **screw** is an inclined plane wrapped around a pole. A wood screw and a jar lid are screws.

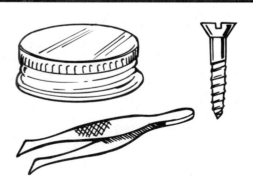

Food Guide Pyramid

The Food Guide Pyramid helps us eat healthy. Color the foods you like.

1. Name two foods in the fruit group.

_____

_____

2. From which group should we eat the most servings?

_____

_____

A Healthy Meal

- one meat or milk
- one or two vegetables
- one fruit
- two or three breads

Draw and color a balanced meal. Use this list and the Food Guide Pyramid on page 50 to help you.

What I Ate

Write everything you eat today in the food shapes. Circle the group you ate the most from.

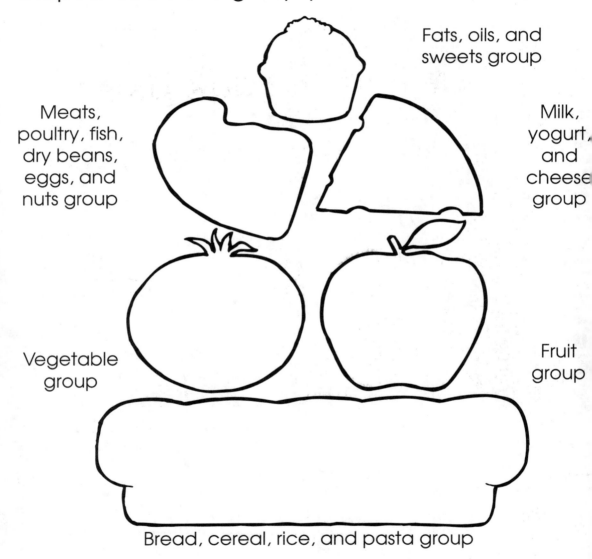

Tomorrow, try to eat more from the groups at the bottom and less from the groups at the top.